GEOTHERMAL POWER

Published by Smart Apple Media
1980 Lookout Drive
North Mankato, Minnesota 56003

Design and Production by EvansDay Design

Photographs: Richard Cummins, Image Finders (Werner Lobert), JLM Visuals (Burton Amundson, Charlie Crangle, Gordon Gilbert, Anthony Irving, Richard Jacobs, Craig Kesselheim), Jim Steinberg

LIBRARY OF CONGRESS CATALOGING-IN-PUBLICATION DATA
Gibson, Diane, 1966–
Geothermal power / by Diane Gibson.
p. cm. — (Sources of energy)
Includes index.
Summary: Defines geothermal energy and explains how it is made and used. Includes a simple experiment.
ISBN 1-887068-76-7
1. Geothermal engineering—Juvenile literature. 2. Geothermal resources—Juvenile literature. [1. Geothermal engineering. 2. Geothermal resources.] I. Title. II. Series.
TJ280.7 .G53 2000
621.44—dc21 99-055892

FIRST EDITION
9 8 7 6 5 4 3 2 1

SOURCES OF ENERGY

geothermalpower

DIANE GIBSON

geothermal power

As rain falls, everything gets wet: trees, cars, houses, and people. The rain pools on the ground, then soaks into it. As the water goes deeper and deeper into the earth, something begins to happen. The temperature inside the earth gets hotter, making the water hotter, too. Eventually, the water becomes so hot that it turns into steam. The steam and heat that rises and escapes through cracks in the ground is called geothermal energy.

DEEP IN THE EARTH

"GEO" MEANS EARTH, or land, and "thermal" means heat. The word "geothermal," then, means "earth heat." To understand where geothermal energy comes from, we need to know what goes on beneath the earth's surface. The earth is made up of three layers. At the center is the core. This part of the earth is extremely hot. Scientists think it is made up of both hard metal and melted rock. Around the core is the mantle, a layer of solid rock that is also very hot. On top of the mantle is the crust, the land that we can see. Over time, cracks form in the mantle and crust. Hot, melted rock called magma sometimes moves up through these cracks. When magma is spit out of **volcanoes**, it is called lava. Magma heats everything around it in the ground, including water. Groundwater is water that lies under the ground in streams or pools. Hot water

THE INSIDE OF THE EARTH IS SO HOT THAT SOME ROCK MELTS INTO A LIQUID FORM CALLED MAGMA.

weighs less than cold water, making heated groundwater rise to the earth's surface. This water may bubble up as a hot spring. If the water is under a lot of **pressure**, it may form a geyser, shooting high into the air every few hours. In other cases, the groundwater touches rock so hot that it leaves the earth as steam. The hole that the steam escapes from is called a fumarole.

Hot springs form anywhere groundwater is heated and rises to the earth's surface.

The temperature inside the earth rises 1° F (.6° C) every 98 feet (30 m) down. Geysers are found in only four countries: Iceland, Indonesia, New Zealand, and the United States.

Volcano lava may be as hot as 2,300° F (1,260° C) when it erupts. But lava is useless in creating geothermal power because it can't be controlled.

ERUPTING LAVA IS EXTREMELY HOT, BUT ITS HEAT CANNOT BE USED TO CREATE GEOTHERMAL POWER.

GEOTHERMAL LOCATIONS

GEOTHERMAL ENERGY CAN be found only in certain places around the world—usually near volcanoes or around deep cracks in the earth. Iceland, Italy, New Zealand, and the United States are some of the countries best able to harness geothermal power. The first power plant to use geothermal heat was built in what is now Larderello, Italy. It has been making **electricity** since 1904. Today, nearly 150 geothermal plants are found all over the world. At a geothermal power plant, hot water may be pumped from the ground and sent directly to a nearby town. Sometimes only steam is brought up from the ground. Steam is what is used to create electrical power.

GEOTHERMAL PLANTS ARE BUILT IN AREAS WHERE THE EARTH PRODUCES HOT WATER OR STEAM.

Geothermal power can often be harnessed in areas of volcanic activity.

GEOTHERMAL PLANTS USE PIPES TO CARRY HEATED WATER OR TO SPREAD COOL WATER ONTO HOT ROCK.

In 1892, more than 200 homes in Boise, Idaho, were heated using hot water directly from the earth. People in Iceland get nearly all of their electricity from geothermal energy.

MAKING ELECTRICITY

WORKERS AT POWER plants drill deep into the ground in areas they know have pockets of hot water or steam. If an area has hot rock but no water, workers will drill there, too. Once a hole is made, water can be poured down to the rock. It then comes back up as steam. The steam is carried by pipes to a machine called a turbine. Sticking out around the turbine are paddles that spin when hit by steam. A pole also sticks out from the center of the turbine. The pole is connected to a generator, the machine that makes electricity. Moving steam spins the turbine, which spins the pole. This in turn spins a large, round **magnet** inside the generator. Around the magnet are thousands of wires twisted into rings. Electricity is created as the magnet spins around inside the rings. The electricity can then be sent to homes and other buildings through cables and wires.

GEOTHERMAL POWER PROBLEMS

GEOTHERMAL HEAT IS a good source of energy, but it can cause some problems. When heated water is taken from the ground, the rock under the ground may become too dry. When this happens, the rock may break or shift suddenly, causing small **earthquakes**. Also, if the water dries up in an area, it could take hundreds or even thousands of years to refill that area with water. To keep these problems from happening, workers plan very carefully before building geothermal power plants.

BECAUSE OF THEIR SHORTAGE OF GROUNDWATER, DESERT REGIONS PRODUCE LITTLE GEOTHERMAL ENERGY.

Rock that is very dry may become unstable, shifting or breaking suddenly.

Heated groundwater that rises through the earth's soil may form bubbling pools called mud spots.

Old Faithful is a geyser in Yellowstone National Park. Every 45 to 90 minutes, it shoots water more than 100 feet (30 m) into the air.

Tourists visit Yellowstone National Park to see its many wonders, including its famous geysers.

THE FUTURE OF GEOTHERMAL POWER

Many electric power plants burn coal or oil to heat water and turn it into steam. But with geothermal energy, nothing is burned. The heat comes naturally from the earth, and the steam comes straight from the ground, so no smoke or dirt is produced. Geothermal power is a good, clean, and dependable source of energy for those areas that can harness it. Power plants can run 24 hours a day throughout the year. They can also make electricity much more cheaply than they can by using other types of fuels. In the future, scientists hope to find even more ways to use the earth's heat.

Much of a geothermal power plant is comprised of pipes and pumps for moving water and steam.

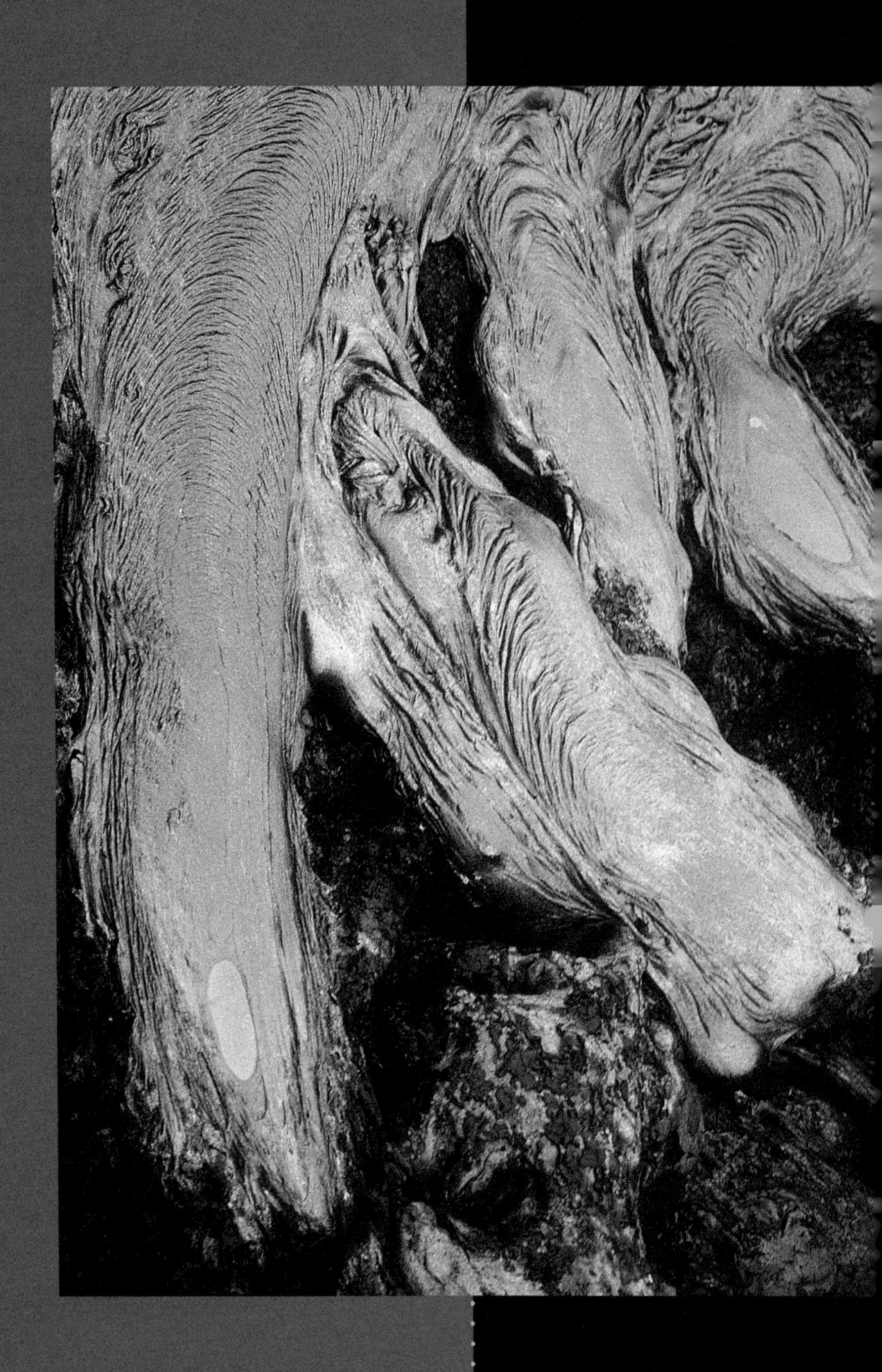

The formation of magma is the first step in the earth's process of producing steam.

Large pumps are used to draw heated water and steam from geothermal wells.

Some geothermal wells produce steam as hot as 700° F (371° C). In the United States, geysers are found only in Yellowstone National Park.

HISTORY

Geothermal heat has been used by people for thousands of years. Native Americans may have used hot springs for cooking and bathing as long as 10,000 years ago. About 2,000 years ago, hot springs were used by the Romans in Italy. They would soak their bodies in the hot water to relax sore or tired muscles. During the 1200s, famed writer Snorri Sturluson of Iceland used hot springs for natural heat. He dug ditches so the water would flow from the springs to his house. Then he put pipes in the walls to carry the water, keeping his house warm during the cold winters.

People have found uses for the warm water of hot springs for thousands of years.

EXPERIMENT

Comparing Fuel Sources

This experiment will show how much cleaner geothermal energy is than fuel-burning sources of power. (For safety, you may need to ask an adult for help.) You will need:

A hot plate or stove top
A pot
Water
Tongs
Two pieces of white paper
A candle
Matches

- Pour some water into the pot and heat it on the stove top or hot plate until it turns to steam. Using the tongs, hold a piece of paper about six inches (15 cm) over the steam. After 30 seconds, put the paper aside.

- Now light the candle. Using the tongs, hold the other piece of paper at the same height over the candle. After 30 seconds, blow the candle out. Look at both pieces of paper. Which is cleaner?

GLOSSARY

Earthquakes ARE SUDDEN SHAKINGS OF THE GROUND CAUSED BY SHIFTING ROCK INSIDE THE EARTH.

Electricity IS A TYPE OF ENERGY USED IN HOMES TO RUN LIGHTS AND APPLIANCES.

A **magnet** IS A PIECE OF METAL THAT ATTRACTS IRON AND STEEL.

Pressure IS A PUSHING FORCE AGAINST AN OBJECT.

Volcanoes ARE MOUNTAINS WITH OPENINGS THAT LET MELTED ROCK AND HOT GASES ESCAPE FROM THE EARTH.

INDEX